Taha Balafrej
Souad Benaich
Rachida Hassikou

Ethnobotanical survey

HF670557

Taha Balafrej
Souad Benaich
Rachida Hassikou

Ethnobotanical survey

on the therapeutic uses of cannabis in the Taounate region of Morocco

ScienciaScripts

Imprint

Any brand names and product names mentioned in this book are subject to trademark, brand or patent protection and are trademarks or registered trademarks of their respective holders. The use of brand names, product names, common names, trade names, product descriptions etc. even without a particular marking in this work is in no way to be construed to mean that such names may be regarded as unrestricted in respect of trademark and brand protection legislation and could thus be used by anyone.

Cover image: www.ingimage.com

This book is a translation from the original published under ISBN 978-620-6-70457-7.

Publisher:
Sciencia Scripts
is a trademark of
Dodo Books Indian Ocean Ltd. and OmniScriptum S.R.L publishing group

120 High Road, East Finchley, London, N2 9ED, United Kingdom
Str. Armeneasca 28/1, office 1, Chisinau MD-2012, Republic of Moldova, Europe
Printed at: see last page
ISBN: 978-620-7-94404-0

Copyright © Taha Balafrej, Souad Benaich, Rachida Hassikou
Copyright © 2024 Dodo Books Indian Ocean Ltd. and OmniScriptum S.R.L publishing group

BALAFREJ Taha

Ethnobotanical survey of the therapeutic uses of cannabis in the Taounate region of Morocco

Summary

Background and objectives: Cannabis has been used for over 4,500 years in various fields, including medicine, to treat a wide range of illnesses. This study aims to conduct an ethnobotanical survey in the province of Taounate to gather information on therapeutic uses by the local population.

Methods: The ethnobotanical study was conducted in the province of Taounate using a questionnaire of semi-structured interviews with randomly selected informants. Data were analyzed using statistical indices and tests such as cultural importance (CI), level of fidelity (NF), relative frequency of citation (RFC), Chi-square (χ^2) test, logistic regression, multiple corresponding analysis (MCA) and non-metric multidimensional positioning (nMDS).

Results: 65.4% of informants use *Cannabis sativa* L for therapeutic purposes, with a CI of around 0.40.

Thus, 11 illnesses were listed for treatment with cannabis, of which "cham" was the most cited, with an NF of 92.9%. The most recommended method of preparation was smoking, and the most common method of administration was inhalation. Results showed a significant correlation between therapeutic use, gender and type of locality. The MCA revealed that male gender, urban areas, high school education and informants aged between 20 and 40 were traits in common among people using cannabis for therapeutic purposes. Secondly, the nMDS showed that the diseases treated were evenly distributed between rural and urban areas.

Conclusion: The present study highlights the importance of the therapeutic use of *Cannabis sativa* L. in the province of Taounate. However, further phytochemical, pharmacological, toxicological and clinical research is needed to investigate its efficacy and safety.

Keywords: *Cannabis sativa* L., ethnobotanical survey, Taounate, therapeutic uses.

Thanks

This research work was carried out in the Taounate region of Morocco in collaboration with the Agence Nationale des Plantes Aromatiques et Médicinales de Taounate, under the scientific co-direction of Madame Rachida Hassikou and Madame Chaimae Rais.

I would like to express my sincere thanks to Madame Rachida Hassikou for her guidance, support and generosity, which have enabled me to progress and improve my knowledge and skills in this learning stage as a doctoral researcher.

I would also like to thank Madame Chaimae Rais for her warm welcome and expertise, which greatly contributed to the success of this work.

Finally, I'd like to thank all those who have contributed in any way for their help, moral support and confidence.

Table of contents

I. Introduction

Medicinal plants represent the oldest and most widespread form of medicine in the world, and continue to be used to this day. Over the past 20 years, increasing attention has been paid to medicinal plants, and numerous ethnobotanical studies have been conducted (Mohammadi *et al.*, 2023, Skalli *et al.*, 2019). It is estimated that 80% of the world's population regularly uses traditional medicine to treat and/or prevent various diseases (Kachmar *et al.*, 2021; Mbuni *et al.*, 2020). In Morocco, there are nearly 600 known and used aromatic and medicinal plants among 4,200 species, whose therapeutic properties are transmitted orally from generation to generation, mainly in rural areas due to the ease of access and affordability of these plants (Jan *et al.*, 2023; Barkaoui *et al.*, 2022).

Cannabis sativa L. is a cosmopolitan dicotyledonous annual plant belonging to the Cannabaceae family, genus Cannabis (Farag & Kayser, 2017). It has been cultivated for over 4,500 years and used in various fields such as medicine, food, textiles and narcotics (Ross, 2005). Interest in cannabis research grew exponentially with the widespread use of the plant for recreational purposes in the 1960s-70s. In the 1980s-90s, the discovery of cannabinoid receptors highlighted their role in several physiological processes to maintain the internal balance of numerous biological functions (Pagano *et al.*, 2022; Pertwee, 2006). Over the last 20 years, ethnobotanical studies have seen a resurgence of interest in various countries around the world, including Morocco (Mohammadi *et al.*, 2023). Numerous therapeutic uses of *Cannabis sativa* L. have been reported in the treatment of various health disorders such as pain, anxiety, nausea, cancer, insomnia and rheumatism (Chaachouay *et al.*, 2022; Amrati *et al.*, 2021; Mouna *et al.*, 2020; Merzouki *et al.*, 1999).

Cannabis cultivation began to proliferate in the 1990s in Taounate province, with a subsequent transfer of traditional knowledge about its various uses to the local population (Blickman, 2017). Over time, Taounate province has become a key region for cannabis cultivation and production (UNODC, 2003). In June 2021, Law 13.21 was enacted. It covers the regulation of the medical, pharmaceutical and industrial use of cannabis, as well as guidelines for its cultivation in specific areas, including Chefchaouen, Taounate and Al Hoceima. This legal framework has facilitated research into the plant (Bachir *et al.*, 2022; Secrétariat général du gouvernement, 2021).

In order to contribute to ethnopharmacology and preserve Moroccan heritage with regard to the traditional medicinal uses of *Cannabis sativa* L., we conducted a survey in the province of Taounate (northern Morocco), where, to our knowledge, few ethnobotanical studies have been carried out concerning cannabis. The main objectives of this work were to explore the extent of cannabis use in this province in terms of the diseases treated by this plant, and to investigate the existence of determining relationships between this local knowledge and various socio-demographic characteristics (age, gender, level of education and type of locality) using statistical tools. Moreover, this work coincides with the implementation of the recent legalization of cannabis cultivation and use in Morocco, where, according to Bellakhdar (1997), cultivation and any use of the plant had been prohibited since 1953. This work highlights the growing importance of cannabis use worldwide, and will help promote research into its pharmaceutical industry.

II. Bibliography

a) Origin and geographical distribution of *Cannabis sativa* L.

Analyses of fossils, pollens and archaeological artefacts have made it possible to roughly trace the origin and dispersal of cannabis through time. Although the geographical origin of cannabis in Eurasia has been established, its exact location remains uncertain (Clarke and Merlin, 2013). Most studies point to a Central Asian origin (Rull, 2022; Ren *et al.*, 2021; Schultes, 1969).

Osterberger *et al*, (2022) suggest that the Yunnan region could be the cradle of cannabis, based on plastome analysis of different cannabis accessions using 8 chloroplast markers. Similarly, McPartland *et al*, (2019) located the center of origin of cannabis on the northeastern Tibetan plateau, near Lake Qinquai (currently Ningxia, China), by analyzing pollen found in 19.6 million-year-old rocks. In contrast, molecular investigations by Zhang *et al*, (2018) suggest a low-latitude region of India as a possible center of origin for cannabis.

Cannabis would then have spread from China to Russia and Europe around 6 million years ago, and to eastern China around 1.2 million years ago, well before the appearance of man. This dispersal is attributed to the last great ice age, during which many plant species, including cannabis, were forced to seek refuge (Clarke and Merlin, 2013; McPartland, 2018). Cannabis cultivation dates back to the Neolithic period in China, as evidenced by archaeological finds dating back 10,000 to 12,000 years in China and Taiwan (Liu *et al.*, 2017). In addition, various uses for cannabis have been documented, including the manufacture of ropes, clothing, sails and vegetable paints found on pottery (Li, 1974).

The plant then spread during the Bronze Age to South Asia through a series of elaborate invasions by Aryan tribes, who greatly contributed to its spread along the Silk Road from Mongolia to the Middle East. Simultaneously, the Scythians, a generally nomadic people of Eurasia, also played a role in the plant's spread to South Asia, the Middle East and Eastern Europe via trade and warfare with Semitic peoples, including the Assyrians, Jews and Egyptians (Warf, 2014).

Cannabis then spread from the Slavic world to Germany following the migration of the Teutonic people and their invasion of Western Europe around the 5th century A.D. (Merlin, 1972).

Cannabis is said to have been introduced into the Maghreb around the 8ème century during the Arab-Muslim conquest (Bellakhdar, 1997), then to have reached Spain at the same time following its invasion by the Moors, a Berber population from North Africa (Warf, 2014). It would then have continued its way to East Africa around the 11th century, passing through Egypt thanks to Arab merchants taking the Red Sea route, before spreading across the rest of the African continent (du Toit, 1976).

Between the 16ème and 18ème centuries, cannabis was introduced to Latin America on several occasions, notably by Spanish colonists (Rubin, 2011). In 1545, the conquistador Pedro Cuadrado, a member of Cortes' army, was the first to introduce the plant to Mexico

City (Clarke and Merlin, 2013). The arrival of British settlers who formed the Jamestown colony in Virginia, in the north of America, enabled the introduction of cannabis and its cultivation for fiber, which then gradually spread throughout the territory (Dr. G.O. Ayenigbara, 2012).

b) Traditional medicinal uses of *Cannabis sativa* L.

Since the Neolithic period, around 12,000 years ago, cannabis has been one of the oldest plants cultivated by man, notably for its fibers (Bonini *et al.*, 2018). It has also been used in various fields such as food, textiles, medicine and recreation (Hartsel *et al.*, 2016). The first document to record its medicinal use is the Pen Ts'ao Ching, dating from 100 to 200 A.D., a compilation of oral transmission from the Chinese emperor Chen Nung who lived 5,000 years ago, considered the father of Chinese agriculture, which prescribed the use of cannabis to treat fatigue, rheumatism, malaria, and more (Touw, 1981).

In India, the use of cannabis dates back to around 1,000 BC (Mikuriya, 1969), with the earliest description found in the Atharva-Veda (3,000-4,000 BC), one of the first sacred Vedic texts to deal with medicine (Sharma, 1977). Numerous other Vedic texts such as the Susrita Samhita describe its therapeutic uses to treat convulsions, fatigue, stimulate appetite, etc. (Pisanti and Bifulco, 2019).

The Assyrian Empire, formed around 2,000 BC, had a good understanding of the therapeutic use of cannabis to treat various ailments such as swelling, bruising, depression, arthritis, kidney stones and gynaecological disorders (R.N., 1997).

In Egypt, as far back as the time of the Pharaohs, cannabis was part of the pharmacopoeia and was administered in a variety of ways, including orally, rectally, vaginally, and by eye or skin fumigation. The first traces of its use date back to the Ramesseum III papyrus, dating from around 1,700 BC, which reports its use to treat eye ailments. Next, the Berlin papyrus, dating from around 1,300 BC, describes the use of cannabis against fever. Finally, the Ebers papyrus, circa 1,550 BC, records its use to induce contractions during childbirth and to treat toenail ailments (Russo, 2007).

Arab medicine, renowned for its advancement, also used cannabis for medicinal purposes. Several Arab travelers and physicians have described its uses, including Al Razi, Sabur Ibn Sahl in his collection "Al Aqrabadhin Al-saghir", Avicenna in his encyclopedia "Avicenna's Canon" (1,000 years A.D.), Ibn Al Baytar in his writings (14ème century), and Marco Polo in his book "The Travels of Marco Polo" (13ème century). Among the treatments described were those for headaches, ophthalmia, uterine pain, general edema, intestinal worms, vomiting, fever, and so on. The plant was also used as an analgesic, anti-inflammatory, antiemetic, anticonvulsant, etc. (Russo, 2007; Mahdizadeh *et al.*, 2015; Nahas, 1982; Hamarneh, 1972).

The ancient Romans and Greeks used cannabis mainly for its fibers in textiles and its seeds in food. Works reporting its medicinal use include Pliny the Elder's "Naturalis historia" (23-79 CE), Dioscorides' "De materia medica" (40-90 CE), and Galen's texts (129-201 CE). Cannabis has been described as a treatment for arthritis and gout by decoction of the roots, to extract worms and parasites from the ears, to relieve wounds, reduce inflammation and oedema, and to treat earache with fresh leaf juice (Crocq, 2020; Pisanti and Bifulco, 2019).

Currently, many communities around the world use the cannabis plant and its derivatives to alleviate and/or treat various illnesses (Bonini *et al.*, 2018). In the Himachal Pradesh district of India, the leaves are used medicinally (Kaushal, 2012). In Morocco's Taounate province, cannabis inflorescence, leaves and seeds are used to treat diabetes, injuries, anxiety, insomnia, cancer, glaucoma, among others (Balafrej *et al.*, 2024). In the Rajshahi district of Bangladesh, cannabis leaves are used to treat insomnia and coughs, while the seeds are used to stimulate sexual function (Nawaz *et al.*, 2009). Cancer patients in Uganda describe using cannabis for vomiting and pain relief (Asiimwe *et al.*, 2023).

c) Regulations

Cannabis sativa is known for its therapeutic properties and its prescription in the treatment of several diseases (Andre *et al.*, 2016). However, the increase in its recreational use in the early 20th^{ème} century led to its restriction in many countries. In 1966, cannabis was classified as a Schedule 1 illicit drug under the Controlled Substances Act, i.e. a substance with a high abuse potential and a serious risk to public health (Svrakic *et al.*, 2012). Prior to this, Egypt was the first to experiment with the banning of cannabis during its occupation by Napoleon, as well as towards the end of the 19ème century, while the United States introduced the Marihuana Tax Act in 1937 to counteract cannabis consumption, and in Morocco in 1953 and 1954 dahirs were promulgated prohibiting the production and consumption of cannabis (Collins, 2020; Bellakhdar, 1997).

Extensive research into the therapeutic potential of cannabis, which began mainly in the 1960s, has led in recent years to the decriminalization and legalization of its use for recreational, medical, industrial or all three purposes in various countries around the world (Budney *et al.*, 2019; Svrakic *et al.*, 2012). California in the USA was the first US state to legalize medical cannabis in 1996, followed by 35 other states in 2020 (Mikos and Kam, 2019). Many other countries have also fully legalized its medicinal use, including Argentina, Canada, Finland, Germany, Italy and Morocco. Other countries such as Belgium, France, Ghana and Japan have restricted its use to specific products and/or under specific conditions (Figure 1) (Decarte, 2019; McGregor *et al.*, 2020; Quarshie and Alagidede, 2020).

In Morocco, the chronology of the cannabis legislation process has gone through several stages (Bennis et *al.*, 2023):

- In 1890, a first attempt was made at the time of Sultan Hassan 1 before the arrival of the French regime in 1906 and in which he allowed its cultivation in 5 douar only of Ketama and Beni khaled;

- In 1906, following the Algeciras conference, the import and export of cannabis was coordinated by foreign diplomatic commissions and the Makhzen;

- In 1916, the French protectorate published a decree legalizing and regulating the cultivation of cannabis;

- In 1932, a decree banned the cultivation of cannabis, with the exception of that grown under the control of the Régie des Tabacs et du Kif in Haouz and Gharb;

- In 1954, the French protectorate issued a decree prohibiting all cannabis cultivation in areas under French domination;

- In 1956, after Independence, there was a ban on cannabis cultivation throughout Morocco, with the exception of historic areas;

- In 1974, the ban on its production and consumption was extended to the whole of the Kingdom;

- In 2021, Morocco adopts a bill to legalize cannabis cultivation.

Figure 1 Legal status of medical cannabis (QGIS map)

d) Taxonomy

The Cannabaceae is a family of annual herbaceous plants and dioecious perennials, rarely monoecious, with alternate or opposite leaves that are often more or less lobed, palmate or compound palmate and stipulate. The flowers are actinomorphic, and wind-pollinated, the fruit is usually an achene or drupe (Xu and Deng, 2017).

This family, named Urticaceae and Moraceae from the mid-18ème century to the mid-20ème century, traditionally contains two genera: Cannabis and Humulus. However, recent phylogenetic analyses have identified a total of 10 genera and over 100 species (The Analytical Chemistry of Cannabis, 2016).

Cannabis has long been the subject of debate over its classification, and still is today. Over time, this classification has been based on many factors, including morphology, environment, chemical composition and genome.

It was first described by Linnaeus as *Cannabis sativa* in 1753, then by J.B. Lamarck in 1785 as *Cannabis indica*, and finally in 1924 by D.E. Janischewsky as *Cannabis ruderalis*, taking into account its morphology, environment and chemical composition (Lapierre et *al.*, 2023). This confusion in terms of classification is subject to controversy as to whether it is poly-typic or monotypic in origin. The poly-typic origin considers the presence of three species: *C. sativa*, *C. indica* and *C. ruderalis*, while the monotypic origin considers only one species: *C. sativa* in which indica and ruderalis and other so-called species are considered varieties (Watts, 2006). It is also suggested that *C. sativa* is the only species with indica, sativa and ruderalis as subspecies.

However, the majority of molecular studies carried out in recent years have concluded in a monotypic origin, recognizing the formal classification of *Cannabis sativa* with as subspecies sativa and indica (McPartland 2018). In addition, a study conducted by Vergara *et al.* (2021) concluded that it is a genetically modified but unique species.

According to the 2016 APG (The Angiosperms Phylogeny Group) classification, the Cannabaceae family can be located as follows:

Kingdom: Plants

Phylum : Phanerogams

Subbranch : Angiosperms

Class: Rosaceae

Subclass: Eurosidae

Order: Rosales

Family: Cannabaceae

Genus: Cannabis L.

Species: sativa

e) Common names

German: Hanf; Haschisch; indischerHanf

English: hemp; marihuana

Arabic: Al-Bhango; Al-Hashish; Al-Qanaap

Chinese: Xian ma; ye ma

Danish: Hemp

Spanish: Canamo; grifa; hachıs; mariguana; marijuana

Finnish: Hamppu

Flemish: Hennep

French: Chanvre; chanvre d'Inde; chanvre indien; chanvrier

Indian: Bhang; charas; ganja

Japanese: Mashinin

Nepalese: Cares; gajiimaa;gaanjaa

Portuguese: Cânhamo; maconha

Russian: Kannabis sativa

Swedish: Porkanchaa (ElSohly et *al.*, 2017)

f) Botanical description

i. Vegetative apparatus

Figure 2 Part of the *Cannabis sativa* stem (Yadav, 2014).

Stem: Stem morphology (Figure 2) remains the same in all *Cannabis sativa* varieties. Stems are generally erect, angular, furrowed, branched, with a woody interior, sometimes hollow in the internodes, and vary in height from 1 to 6 m (1 to 3 m on average). Branches are either opposite or alternate. However, certain differences can be observed between male and female plants: male plants are around 10 to 15% taller than females, which are wider (Bouloc, 2006).

Leaves: The leaves are formed by a long petiole bearing 5 to 7 green, lanceolate, palmate, unequal, elongated and toothed leaflets (Figure 3). The phyllotaxy of this species evolves all the time, at first appearing as an opposite phyllotaxy then becoming alternate (Farag and Kayser, 2017).

Figure 3 Abaxial (left) and adaxial (right) faces of a *Cannabis sativa* leaf (Yadav, 2014).

Roots: Hemp's root system consists of a branched taproot (figure 4), reaching depths of 30 to 60 cm and up to 2.5 m in loose soils (Farag and Kayser, 2017). Male plants are distinguished from females by a shorter vegetative period and therefore a less developed root system (Bouloc, 2006).

Figure 4 *Cannabis sativa* root apparatus (Farinon *et al.*, 2020).

ii. Reproductive system

Cannabis sativa is a dioecious plant with an average life cycle of 4 to 6 months. It is mainly allogamous, so its pollen is spread by the wind, and it reproduces exclusively by seed.

Once the seeds have germinated, it is difficult, if not impossible, to determine the sex of the plant until it reaches maturity. This is because a number of factors can modify the nature of the sex, mainly environmental factors and hormones. Subsequently, progress towards the flowering stage is induced by short photoperiods on consecutive days (10 to 12 hours of darkness).

Figure 5 Flowers of *Cannabis sativa*. L, male (A) and female (B) (Bonini *et al*., 2018).

Flowers: Male and female flowers differ; female flowers are characterized by an absence of petals, 2 yellow/whitish stigmas with an ovary enclosing a single ovule, an inflorescence in compact cymes mingled with foliaceous bracts, and inserted in pairs at these pistils supported on the stem by small leaves (bracts). Male flowers, on the other hand, are grouped in panicles, with 5 green or yellow petals, 5 pale green sepals and 5 hanging stamens with filaments and slender stamens (figure 5).

Seeds: *Cannabis sativa* seeds are ovoid or spherical in shape, 3 to 5 mm long (figure 6). They contain an embryo which occupies the entire volume and is covered by its own envelope (testa) and pericarp. 2 cotyledons rich in reserves are also present.

g) Ecology

Figure 6 *Cannabis sativa* seeds (Farinon *et al.*, 2020).

The morphological, biological and pharmaco-chemical characteristics of the *Cannabis sativa* species depend on growing conditions, altitude, temperature, humidity, light, and the types of fertilizers used (Stambouli et *al.*, 2005). Optimal growth of this herb requires a temperature averaging 14.3°C, a pH of 6.5, an average rainfall of 950 mm, good light, a humid climate or moist soil with good drainage and a soil rich in mineral nutrients (mainly nitrogen in large quantities) and organic matter (CABI, 2021).

III. Materials and methods

a) Study area

Taounate province is part of the Fès-Meknès region in Morocco, located between 34°32'09" north and 4°38'24" west. It is one of the 12 regions created during the territorial reorganization in 2015, with Fès as its capital (Bulletin Officiel, 2015). Taounate is bordered by the provinces of Chefchaouen and Al Hoceima to the north, the region of Fès to the south, the province of Taza to the east and the province of Sidi Kacem to the west (Figure 7). Covering a total area of 5,585 km2, Taounate province comprises 49 communes, including 5 urban areas and 44 rural areas, as well as almost 1,600 villages (Figure 7). The region lies in a Mediterranean climate zone characterized by cold, wet winters (with an average of 5°C) and variable rainfall, with northern areas receiving up to 1,800 mm of precipitation per year (Agence urbaine de Taza, 2017).

According to the UNODC survey on cannabis in Morocco (2004), the economy of Taounate province is based on cereal farming, olive growing and livestock breeding. Moreover, since the end of the last century, cannabis cultivation has grown exponentially, mainly among young people. These data provide a general overview of Taounate province and highlight its importance in terms of cannabis cultivation and its potential for scientific research within the national regulatory and legislative framework established since June 2021

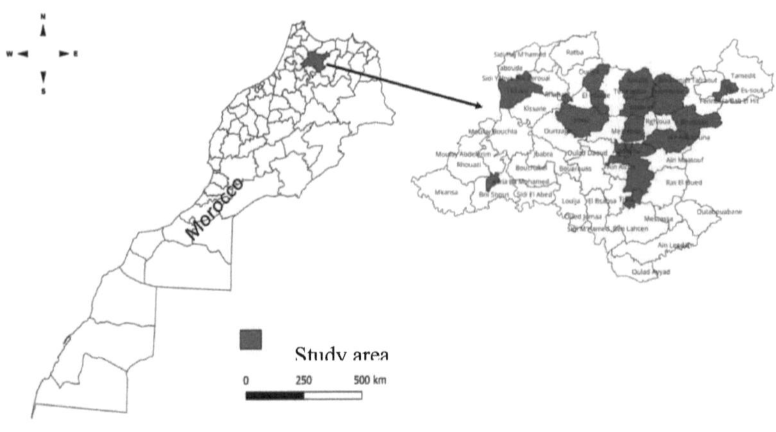

Figure 7 Study areas in Taounate province (Map modified using QGIS software).

b) Data collection

The ethnobotanical survey of *Cannabis sativa* L. was conducted from December 15, 2022 to January 15, 2023 in 15 municipalities, including 5 urban areas (Taounate, Tissa, Karia Ba Mohammed, Ghafsai and Thar Es-Souk) and 10 rural areas (Bouhouda, Bni Oulid, Ain Mediouna, Khlalfa, Tafrant, Zrizer, Sidi El Makhfi, Ain Aicha, Galaz and Fericha). Data were collected through random face-to-face interviews at weekly markets, group discussions in the field and using a semi-structured questionnaire. The selection of municipalities was based on interviews with the local population, existing literature and the intention to reach a large number of indigenous people throughout the province. A total of 422 participants were interviewed after obtaining their verbal consent, using a questionnaire divided into two parts: the first part dealt with socio-demographic characteristics (age, gender, level of education, occupation, type of municipality and name of municipality), while the second part included questions on plant use, methods of use, vernacular names, plant parts used, diseases treated and dosage.

c) Quantitative ethnobotanical data

i. Relative Citation Frequency (RCF)

The Relative Frequency of Citation index has been used to assess the importance of plant use within a specific population (Tardío & Pardo-de-Santayana, 2008). It is measured on a scale of 0 to 1, where values close to 0 indicate a lower frequency of citation and, consequently, less cultural importance. On the other hand, values close to 1 indicate a higher frequency of citation and, consequently, greater cultural importance within the population. The FRC index is calculated using the following formula:

$$FRC = FC/N$$

FC: Number of informants mentioning the use of species X; N: Total number of informants.

ii. Index of Cultural Importance

The Cultural Importance Index (CI) is a valuable tool for measuring variations in knowledge between different communities and assessing the importance of a plant in a specific use category (Asseh *et al.*, 2019). The values obtained using the IC index provide a quantified measure of the importance of each use category in a particular environment. The formula used to calculate this index is as follows:

$$IC = UR/N$$

UR: The number of informants using a given species for a specific use category; N: Total number of informants.

iii. Loyalty level

The Fidelity Level (FL) was developed by Friedman *et al*, (1986), and is a useful tool for measuring the importance of a plant's specific uses within a particular use category (Ong and Kim, 2014). The values obtained with this index range from 0% to 100%. The index is defined using the following formula:

NF = (Ni/N) *100

Ni: Number of informants citing specific use; N: Total number of informants.

d) Data analysis

All data collected during this ethnobotanical survey were transcribed and a descriptive analysis was carried out using Sphinx Plus 2-V5. We then used statistical tests to explore and obtain information on the factors influencing the use of *Cannabis sativa* L. for therapeutic purposes.

Bivariate analysis (Chi-square): This method was used to examine associations between the use of *Cannabis sativa* L. for therapeutic purposes and each of the socio-demographic variables.

Multivariate Logistic Regression Analysis: The aim of this method was to determine how sociodemographic factors combined to influence the therapeutic use of *Cannabis sativa* L. Its contribution lay in highlighting the most significant factors affecting the therapeutic use of *Cannabis sativa* L. while taking into account several variables simultaneously.

Multiple Correspondence Analysis (MCA): MCA was used to understand the typology of users of *Cannabis sativa* L. for therapeutic purposes, taking into account all socio-demographic variables. It enabled us to create a graphical representation of the relationships between variables, and to reveal associations and patterns not immediately evident in the raw data.

Non-Metric Multidimensional Analysis (nMDS): The aim of this analysis was to understand the relationships between diseases treated with *Cannabis sativa* L. and geographical location. It enabled us to visualize how different types of illness treated with *Cannabis sativa* L. relate to specific geographical regions, offering a spatial perspective on patterns of therapeutic cannabis use. We also tested whether rural and urban localities differed in diseases treated with *Cannabis sativa* L. using a permutable multivariate analysis of variance (PERMANOVA) with a significance level of 0.05 and 999 permutations and with the adonis2 function in the "vegan" R package (Oksanen *et al.*, 2012).

IV. Results

a) Socio-demographic characteristics of informants

The survey involved face-to-face interviews with a total of 422 individuals. Of these, 65.4% (276 out of 422, comprising 221 men and 55 women) reported using *Cannabis sativa* L. for various purposes. Table 1 provides an overview of the socio-demographic characteristics of participants using *Cannabis sativa* L.

The age of participants using the plant ranged from 15 to 67. The most common age groups were between 20 and 40 (44.2%) and between 40 and 60 (42.8%). In terms of level of education, those with no schooling (16.3%) and primary education (40.6%) accounted for over half the participants. The percentage then drops significantly among participants with secondary education (34.8%) and university education (8.3%). Furthermore, 56.9% of participants came from rural areas and 43.1% from urban areas.

Variable	Subgroup	Number	Percentage
Gender	Men	221	77%
	Woman	55	23%
Age	<20	5	1.8 %
	20-40	122	44.2 %
	40-60	118	42.8 %
	>60	31	11.2 %
Education level	Out of school	45	16.3 %
	Primary	112	40.6 %
	Secondary	96	34.8 %
	University	23	8.3 %
Type of location	Urban	119	43.1 %
	Rural	157	56.9 %

Table 1 Socio-demographic characteristics of informants using *Cannabis sativa* L.

b) Influence of sociodemographic factors on *therapeutic* use Cannabis sativa L.

The results of the Chi-square test analysis revealed that gender and type of locality both had a significant association effect on the therapeutic use of *Cannabis sativa* L., as shown by the respective results (p = 3.545 e-06) and (p = 4.843 e-06) (Table 2).

Distribution	Chi-Carré	*p* value

		Therapeutic use	Non-therapeutic use	$()\chi^2$	
Age	<20	3	16	6.67	0.1538
	20-40	76	113		
	40-60	73	98		
	>60	16	27		
Gender	Men	149	176	20.41	3,545 e $^{*-06}$
	Woman	19	78		
Education level	Out of school	27	44	1.30	0.7291
	Primary	68	99		
	Secondary	63	89		
	University	10	22		
Type of location	Rural	82	180	20.89	4,843 e $^{*-06}$
	Urban	86	74		

Table 2 Influence of sociodemographic factors on the therapeutic use *of Cannabis sativa* L.

Furthermore, the results of the multivariate logistic regression analysis revealed that women had significantly lower odds compared to men, who served as the reference category, as indicated by an odds ratio of 0.2 (95% CI: [0.14, 0.44]) and a p-value less than 0.01, signifying a highly significant difference. Furthermore, our results indicated a significant association between living in urban areas and the therapeutic use of *Cannabis sativa* L. (odds ratio 2.69, p-value < 0.001) (Figure 8).

Variable		N	Odds ratio		p
Sex	Men	325		Reference	
	Women	97		0.25 (0.14, 0.44)	<0.001
Type.of.locality	Rural	262		Reference	
	Urban	160		2.69 (1.76, 4.16)	<0.001
Educational.level	Primary	167		Reference	
	Secondary	152		1.06 (0.65, 1.73)	0.81
	University	32		0.68 (0.28, 1.59)	0.39
	Unschooled	71		0.77 (0.41, 1.41)	0.40
Age	<20	19		Reference	
	>60	43		2.97 (0.77, 14.90)	0.14
	20-40	189		3.31 (1.01, 14.94)	0.07
	40-60	171		3.70 (1.13, 16.73)	0.05

Figure 8 Multiple logistic regression analysis of socio-demographic factors and therapeutic use of *Cannabis sativa* L.

Analysis of sociodemographic variables and respondents' therapeutic/non-therapeutic use using multiple correspondence analysis (MCA)
Multiple correspondence analysis (MCA) of the different variables showed that the two dimensions alone absorb 29.7% of the total inertia (Figure 9). This allows us to classify all the modalities into three distinct groups.

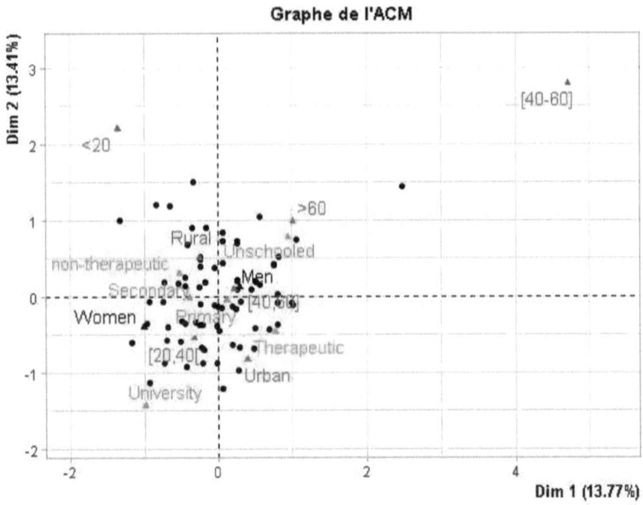

Figure 9 Visualization of the Multiple Correspondence Analysis (MCA)

Group 1 is made up of people living in rural areas who do not use *Cannabis sativa* L. for therapeutic purposes. This group is mainly made up of women under 20 with limited formal education. In addition, it includes individuals over 60, mainly men living in urban areas, who use *Cannabis sativa* L. for therapeutic purposes.

The results of Group 2 highlight notable trends in the use of *Cannabis sativa* L. for therapeutic purposes. The analysis reveals that the majority of users are male and tend to live in urban areas. The age bracket most frequently associated with therapeutic use of *Cannabis sativa* L. is between 40 and 60, followed by the 20-40 age group. On the other hand, individuals under 20 and over 60 are less represented in the population of therapeutic users of *Cannabis sativa* L. In addition, a significant proportion of women belong to the category of non-users.

Group 3 reveals significant trends in age, education and gender. There is a notable representation of individuals over 60 and those with no formal education, with both men and women included in these categories. On the other hand, participants with secondary education are also part of the study, demonstrating a substantial distribution in the 40-60 and 20-40 age groups.

c) Common names

In the region studied, informants reported that "l'kîf" was mainly used for psychotropic purposes and "qanneb" for textile use as vernacular names for the species *Cannabis sativa* L.

d) Use categories

Our survey revealed four categories of cannabis applications reported by participants. Of these, the most frequently mentioned category was cosmetics (used for hair, nails and

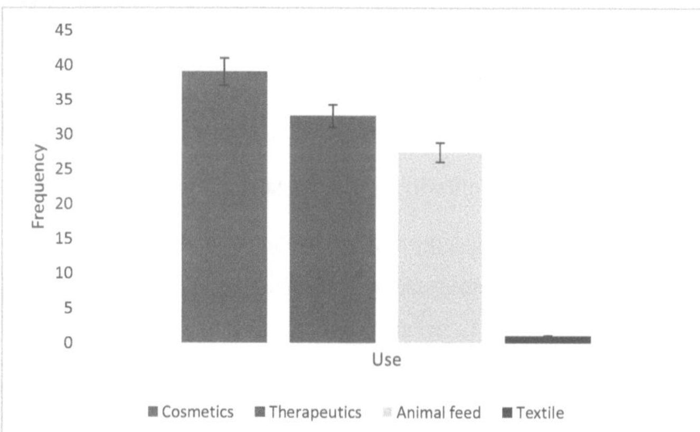

skin), accounting for 39%. Next came the therapeutic category (used to treat human ailments), with 32.6%, followed by the animal feed category with 27.4% and the textile category (used to make ropes), with just 1% (Figure 10).

Figure 10 Categories of *Cannabis sativa* L. use in Taounate province

e) Diseases treated by *Cannabis sativa* L.

The survey revealed that cannabis was used as a treatment for 11 different illnesses. These treatments were mainly based on recommendations from acquaintances, family members, herbalists or by themselves. The results showed that 156 respondents (50.8%) mentioned the use of cannabis to treat Cham. A little-documented illness, it affects children aged between 6 and 12 months and is characterized by symptoms such as head swelling, diarrhea and incessant crying. According to the participants, this illness is generally caused by a strong smell or by magic. In addition, 34 (11.1%) respondents reported its use for diabetes, 25 (8.1%) for injuries, 24 (7.8%) for pain, 21 (6.8%) for anxiety, 17 (5,5%) for insomnia, 9 (2.9%) for cancer, 8 (2.9%) for headaches, 6 (2%) for stomach disorders, 5 (1.6%) for dizziness and 2 (0.7%) for glaucoma (Figure 11).

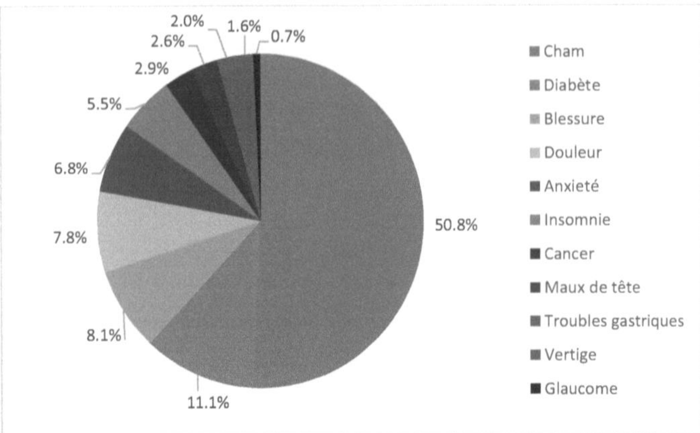

Figure 11 Diseases treated by informants using *Cannabis sativa* L. as a medicinal plant.

f) Analysis of sample dissimilarity by disease category in rural and urban locations

The metric-free multidimensionality (nMDS) plot in Figure 12B reveals distinct patterns that are key to our understanding of the relationships between disease types treated with *Cannabis sativa* L. and geographic locality.

The Shepard diagram (reveals a significant correlation between observed distances and distances calculated by ordination, with a coefficient of determination R2 exceeding 0.99.

This high correlation attests to the reliability of the NMDS adjustment. What's more, the low stress value, measured at just 0.07, also confirms the high quality of the NMDS fit (see **Figure 12A**). This combination of results underlines the relevance of the two-dimensional representation of object similarities in the analysis, as well as the robustness of the method used for this study.

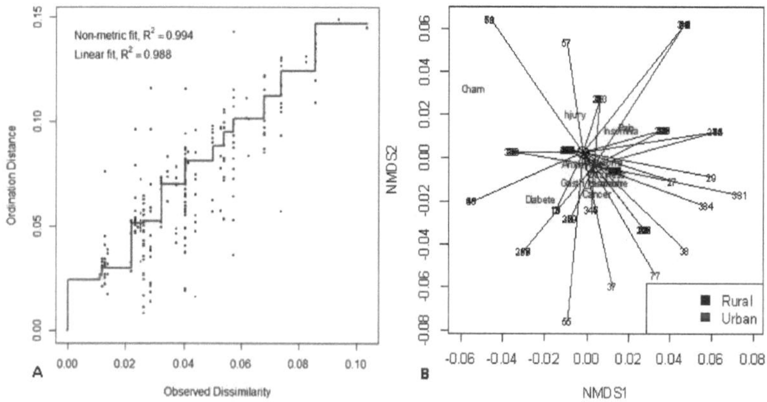

Figure 12 (A) Shepard stress diagram. Relationship between NMDS ordination distance and original observed distance. (B) Non-metric multidimensional scaling (NMDS) plot based on Bray-Curtis.

The dissimilarities were clearly evident in the non-metric multidimensional scaling (NMDS) plot, as shown in Figure 6B. Samples showed no distinct separation according to *Cannabis sativa* L. treated diseases or type of locality. In fact, the points representing diseases treated with Cannabis sativa L. were uniformly scattered across the graph, far from the centroid of the ellipse that delineates the two types of locality, rural and urban. These results suggest the absence of any significant relationship between the diseases treated and the type of geographical locality in the NMDS analysis, indicating the absence of any significant clustering or trend in the distribution of samples according to these factors.

In light of the visually suggestive results of the nMDS graph, the imperative need for rigorous statistical validation of our hypothesis becomes clear. To this end, our PERMANOVA analysis revealed that there were no significant differences in illnesses treated with *Cannabis sativa* L. between the rural and urban localities studied ($F = 2.564$; $p = 0.055$).

g) Plant part(s) used, methods of preparation and administration

The most commonly used plant parts were leaves and flowers (46.4%), followed by seeds (7.2%). Among informants, 86.5% of plant parts used were dried and 13.5% were fresh (Figure 13).

In the current study, smoking was the most common preparation method (83.1%), followed by the use of vegetable oil (13.9%) and powder (3%) (Figure 14). For example, smoking *Cannabis sativa* L. leaves and/or flowers has been suggested for the treatment of Cham (Table 3). The most common route of administration was inhalation (83.1%), followed by topical oil application (13.9%) and oral administration (3%).

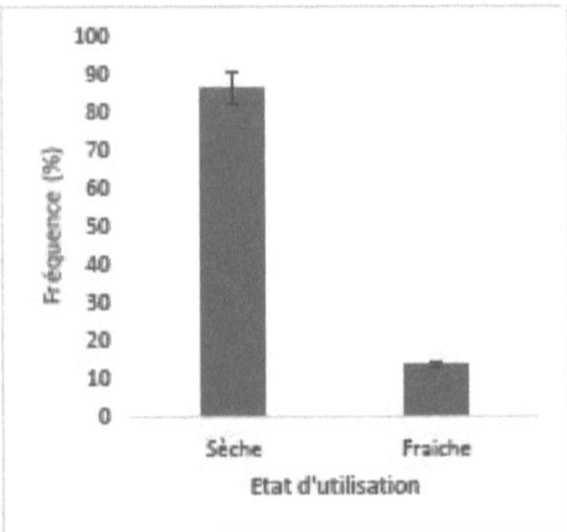

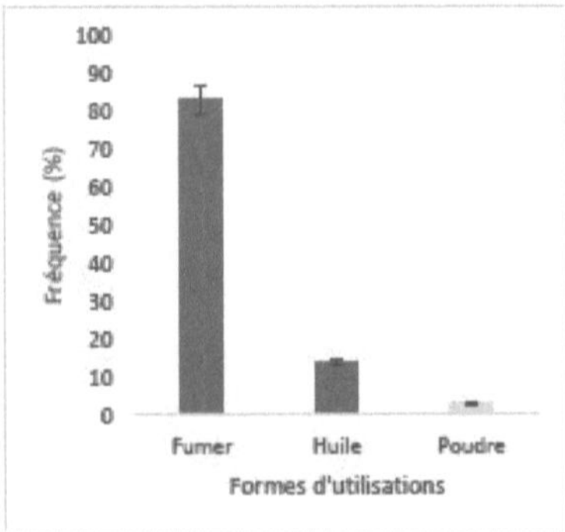

Figure 13 Status of plant parts used by informants

Figure 14 Consumption methods reported by informants.

Table 3 *Cannabis sativa* L. plant parts used by local people in Taounate province.

Plant parts	Method of preparation/administration	Treated conditions
Leaves/Flowers	Smoked with or without *Nicotiana tabacum* leaves using a sebsi with a narrow clay bowl called "Shqaf"	Cham; diabetes; pain; anxiety; insomnia; cancer; headache; vertigo; glaucoma.

	representing approximately 25 mg; inhaled.	
Leaves/Flowers	A small handful of powder mixed with pure honey and *Trigonella foenumaraecum*; taken orally.	Gastrointestinal diseases
Seeds	Oil mixed with or without olive oil; topical application	Injuries

h) Dosage

In this survey, the majority of informants used *Cannabis sativa* L. as a medicinal plant without a specific dose. In fact, results showed that smoking *Cannabis sativa* L. leaves mixed with *Nicotiana tabacum* leaves and blowing them on the patient 3 to 5 times cures cham. In addition, one or two spoonfuls are used to treat gastric disorders. The duration of use of *Cannabis sativa* L. as a medicinal plant by the local population was not specific. According to the population surveyed, no adverse effects were reported.

i) Quantitative analysis

In this study, RFC values for Cannabis sativa L. in the province of Taounate were 0.65. This value shows that over 50% of the local population use the plant for therapeutic, cosmetic, food or textile purposes. Among the categories of use, cosmetics had the highest IC value of 0.48, followed by therapeutics with an IC of 0.40, while animal feed had an IC of 0.33 and textiles an IC of 0.01 (Table 4).

Category of use	Cultural Importance Index (CI)
Cosmetics	0.48
Therapeutics	0.40
Animal feed	0.33
Textile	0.01

Table 4 Cultural Importance Index of *Cannabis sativa* L. in the different localities visited.

Our results show that the Cannabis plant is reported to treat 11 different disease categories. Among these categories, the treatment of Cham was associated with a very high level of fidelity, with a remarkable value of 92.9%. Diabetes was the next most notable category, with a fidelity level of 20.2%, followed by injuries at 14.9% and pain at 14.3% (Table 5). These results highlight the perceived efficacy of the cannabis plant in the treatment of these particular illnesses according to survey respondents.

Table 5 Percentage fidelity of diseases treated with *Cannabis sativa* L.

Diseases	Number of quotes	Fidelity level (NF)
Cham	156	92.9 %
Diabetes	34	20.2 %
Injuries	25	14.9 %
Pain	24	14.3 %
Anxiety	21	12.5 %
Insomnia	17	10.1 %
Cancer	9	5.3 %
Headaches	8	4.8 %
Gastrointestinal disorders	6	3.6 %
Vertigo	5	3%
Glaucoma	2	1.2 %

V. Discussion

In our survey, the socio-demographic results showed that the age of all participants ranged from 15 to 67, and the most represented group was those aged 20 to 40 (44.2%), followed by those aged 40 to 60 (42.8%). Older people seem to have better knowledge than other age groups, as reported by Bouarfa *et al.*, (2020) in Taounate province and Skalli *et al.*, (2019), as well as Barkaoui *et al.*, (2022) in various provinces of Morocco. A significant association between gender and the therapeutic use of cannabis was observed, with men showing greater knowledge of cannabis than women, as mentioned by Bouarfa *et al.*, (2020) and Merzouki and Molero Mesa (1999), in contrast to what can be found in most surveys of other species, as mentioned by Sekkat *et al.*, (2023). This may be due to the fact that men use this plant more often, mainly for psychotropic purposes, as well as other uses.

In terms of level of education, the data showed that 50% of informants were unschooled or had a primary level of education. This means that cannabis use is more common among those with no schooling or a primary level of education, which may suggest that they have more expertise in its uses. These results are similar to those reported by Kachmar *et al*, (2021) in the provincial region of Taza, and Benkhaira *et al*, (2021) in the same province as our study. Moreover, no significant association was observed between informants' level of education and their therapeutic knowledge of the plant. Furthermore, the study highlights the emphatic use of medicinal plants in rural areas compared to urban areas, with a prevalence rate of 56.9%. However, urban areas reported a significant relationship with the therapeutic use of cannabis. This can be attributed to the rural exodus that began in the 1970s and continues today, with 152,000 migrants per year according to the HCP (2023), taking their knowledge with them from rural to urban areas.

In traditional uses, *Cannabis sativa* L. could be used in various sectors such as tool-making (clothing, rope and paper), cosmetics, medicines, human and animal food (Hourfane *et al.*, 2023; Chaachouay *et al.*, 2022). According to Bellakhdar (1997), Moroccans have been using cannabis since 1800, mainly for its psychoactive and therapeutic properties. Our study showed that cosmetics and medicines were the main sectors using *Cannabis sativa* L. in Taounate province, with over 71.6%. In addition, the Chi-square test showed a significant correlation between the therapeutic use of *Cannabis sativa* L. and the variables of gender and type of locality. Multivariate analysis indicated that men and individuals living in urban localities showed a highly significant association with therapeutic cannabis use compared with women and individuals living in rural localities. These results are confirmed by Dalli *et al*, (2023) and Bellakhdar (1997).

The RFC value for cannabis in the province of Taounate was 0.65, meaning that over 50% of the local population have knowledge of the traditional uses of cannabis. Among these different uses, therapeutic use was presented by a cultural index value of 40%. The plant's therapeutic properties have been experimentally demonstrated by several in vitro and in vivo studies (Haddou *et al.*, 2023; Kim *et al.*, 2023; Breijyeh *et al.*, 2021). Our survey reported 11 diseases treated with cannabis. Of these, cham was the disease most cited as treated by cannabis in the local population, as indicated by their notably high frequency of use and FL value. Moreover, no significant relationship between treated diseases and geographical locality was observed. In another ethnobotanical study conducted by Bouarfa *et al* (2020) in the province of Taounate, cham also topped the list. Other studies carried out by Amrati *et al*, (2021) among communities living in mountainous areas of the Fès-Meknès region, and by Benali *et al*, (2017) in the province of Guercif (north-east Morocco) reported that this plant is used to treat cancer. In the Moulay Yacoub region and Taounate province, the plant is also consumed to treat skin diseases (El Khomsi *et al.*, 2022; El-Hilaly *et al.*, 2003).

The use of leaves and flowers was reported as the most widely used parts of *Cannabis sativa* L. (46.4%), followed by seeds (7.2%). The use of leaves and flowers can be explained by the fact that their composition features a variety of secondary metabolites such as cannabinoids, terpenoids and flavonoids (Judžentienė *et al.*, 2023; Chaachouay *et al.*, 2022). Indeed, the use of leaves could be explained by the ease of harvesting and the simplicity of preparing herbal remedies (Benkhaira *et al.*, 2021; El-Assri *et al.*, 2021). In terms of preparation method, smoking was the most commonly used by informants due to their greater knowledge of this method. This observation is in line with Balant *et al*, (2021), who indicate that smoking is the most frequently used way of consuming cannabis.

Furthermore, the administration method most used by informants in our survey was inhalation, as mentioned by Bouarfa *et al.*, (2020). This method is preferred by informants due to the rapid absorption of active ingredients by inhalation, resulting in a relatively quick onset of effects within minutes (Bridgeman & Abazia, 2017). Combustion of cannabis increases the concentration of inhaled Δ9-tetrahydrocannabinol (THC) through decarboxylation of Δ9-tetrahydrocannabinolic acid (THCA) (Gigopulu *et al.*, 2022). In contrast, orally administered cannabis does not produce an effect for 30 minutes to 2 hours, due to absorption of THC by the intestine and transport to the liver, according to the National Academy of Sciences report on the health effects of cannabis and cannabinoids.

The duration of *Cannabis sativa* L. treatment among informants was not specified, indicating that it can vary from one day to several weeks or even years. In addition, the dose of cannabis was not specified by the informants. Unfortunately, unspecified doses of medicinal plants can lead to adverse effects, as some plant species become toxic if misused (El-Assri *et al.*, 2021). The efficacy of cannabis in treating the diseases mentioned needs to be further investigated through research studies. Such studies would provide further evidence to confirm the therapeutic potential of this plant for various health conditions.

This study was carried out only in the Taounate region and for a short period. A longer period adding other localities and regions complying with law 13.21 would be added to obtain better results. The study highlights the potential of *Cannabis sativa* L. as a source of medicines for various diseases, warranting further preclinical and clinical investigation. However, it is important to recognize the challenges encountered in data collection due to the reluctance of many individuals to share their knowledge. This reluctance may stem from fear of the authorities or confusion as to the researchers' intentions.

VI. Conclusion

The present survey represents one of the few studies focusing on ethno-pharmacological knowledge held by informants in the province of Taounate using *Cannabis sativa* L. as a medicinal plant. These participants provided detailed information on cannabis uses, vernacular names, diseases treated, plant parts used, preparation and administration methods.

The results revealed that 32.6% of informants used this plant therapeutically to treat 11 different illnesses. With regard to the relationship between socio-demographic variables and therapeutic use by the population, common traits were observed: men, aged between 20 and 40, with secondary education and living in urban areas.

Following the legalization of cannabis for industrial, pharmaceutical and cosmetic purposes, which had been banned from 1953 to 2021, we encountered challenges in this survey. Some participants were reluctant to respond due to concerns about the authorities. In addition, further surveys are needed to expand the scope to other regions, as permitted by Law 13.21.

VII. References

Taza Urban Agency. 2017. Taounate city development plan. https://www.autaza.ma/DocumentsAUT/DocumentsUrbanisme/TAOUNATE/note-presentation-.pdf. (Accessed 06/11/2023).

Amrati FEZ, Bourhia M, Slighoua M, Salamatullah AM, Alzahrani A, Ullah R, Bari A, Bousta D. 2021. Traditional medicinal knowledge of plants used for cancer treatment by communities of mountainous areas of Fez-Meknes-Morocco. Saudi Pharmaceutical Journal 29 (10): 1185-1204. https://doi.org/10.1016/j.jsps.2021.09.005.

Andre, Christelle M., Jean-Francois Hausman, and Gea Guerriero. 2016. "Cannabis Sativa: The Plant of the Thousand and One Molecules". *Frontiers in Plant Science* 7 (February). https://doi.org/10.3389/fpls.2016.00019.

Asiimwe, John Baptist, Prakash B. Nagendrappa, Esther C. Atukunda, Ivan Kahwa, Lina S. Mathew Alonga, Clement O. Ajayi, Casim U. Tolo, Patrick E. Ogwang, and Maud M. Kamatenesi. 2023. "Ethnobotanical Survey of Medicinal Plants Used in the Management of Cancer in Uganda". *Journal of Herbal Medicine* 42 (December): 100808. https://doi.org/10.1016/j.hermed.2023.100808.

Asseh EE, Yao K, Ake-Assi E. 2019. Diversity and Ethnobotanical Knowledge of Species of the Family Acanthaceae of the Partial Nature Reserve of Dahliafleur, Côte d'Ivoire. European Scientific Journal 15. https://doi.org/10.19044/esj.2019.v15n9p444.

Bachir F, Eddouks M, Arahou M, Fekhaouui M. 2022. Origin, Early History, Cultivation, and Characteristics of the Traditional Varieties of Moroccan *Cannabis Sativa L.* Cannabis and Cannabinoid Research 7 (5): 603-15. https://doi.org/10.1089/can.2021.0020.

Balafrej, T., Skalli, S., Benaich, S., Bessi, A., El Mekkaoui, A., Boukour, B., ... & Hassikou, R. (2024). An ethnobotanical survey on the therapeutic use of Cannabis sativa L. in the province of Taounate, Morocco. *Ethnobotany Research and Applications*, *28*, 1-14.

Balant M, Gras A, Ruz M, Valles J, Vitales D, Garnatje T. 2021. Traditional uses of Cannabis: An analysis of the CANNUSE database. Journal of Ethnopharmacology 279: 114362. https://doi.org/10.1016/j.jep.2021.114362.

Benali T, Khabbach A, Ennabili A, Hammani K. 2017. Ethnopharmacological prospecting of medicinal plants from the Province of Guercif (NE of Morocco). Moroccan Journal of Biology 14(14): 1-14.

Benkhaira N, Ech-Chibani N, Fikri-Benbrahim K. 2021. Ethnobotanical Survey on the Medicinal Usage of Two Common Medicinal Plants in Taounate Region: Artemisia Herba-Alba Asso and Ormenis Mixta (L.) Dumort. Ethnobotany Research and Applications 22: 1-19.

BENNIS, Soukaina, Majda HEDDATE, and Amal Ait Haj Said. 2023. "Current status of cannabis regulation in Morocco." *World Journal of Biology Pharmacy and Health Sciences* 15 (August): 151-56. https://doi.org/10.30574/wjbphs.2023.15.2.0350.

Blickman T. 2017. Morocco and Cannabis. Reduction, Containment or Acceptance. "Drug Policy Briefing. Pobrane z: https://www. researchgate.

net/publication/316104129_Morocco_and_Cannabis_Reduction_containment_or_acceptance. Breijyeh, Z, Jubeh, B, Bufo SA, Karaman R, Scrano L. 2021. Cannabis: A Toxin-Producing Plant with Potential Therapeutic Uses. Toxins 13 (2): 117. https://doi.org/10.3390/toxins13020117.

Bonini, Sara Anna, Marika Premoli, Simone Tambaro, Amit Kumar, Giuseppina Maccarinelli, Maurizio Memo, and Andrea Mastinu. 2018. "Cannabis Sativa: A Comprehensive Ethnopharmacological Review of a Medicinal Plant with a Long History". *Journal of Ethnopharmacology* 227 (December): 300-315. https://doi.org/10.1016/j.jep.2018.09.004.

Bouarfa M, Lebtar S, Boukhira S, Bousta D. 2020. An Ethnobotanical and Ethnopharmacological Survey of *Cannabis sativa* of Taounate Region in Northern Morocco. International Journal of Pharmaceutical Sciences Review and Research. Vol. 64. https://doi.org/10.47583/ijpsrr.2020.v64i02.019.

Bouloc P. Industrial hemp: production and uses. Paris: France agricole. 2006.

Budney, Alan J., Michael J. Sofis, and Jacob T. Borodovsky. 2019. "An update on cannabis use disorder with comment on the impact of policy related to therapeutic and recreational cannabis use." *European archives of psychiatry and clinical neuroscience* 269 (1): 73-86. https://doi.org/10.1007/s00406-018-0976-1.

Bulletin Officiel. 2015. Décret n°2-15-40 du 1er Joumada I 1436 fixant le nombre de régions, leurs dénominations, leurs chefs-lieux ainsi que les préfectures et provinces qui les composent .https://knowledge-uclga.org/IMG/pdf/decret_no2-15-40_du_1er_joumada_i_1436_20_fevrier_2015_fixant_le_nombre_des_regions_leurs_denominations_leurs_chefs-lieux_ainsi_que_les_prefectures_et_provinces_qui_les_composent_bo_no6340_du_5_mars_2015_.pdf. (accessed 15/03/2023).

Bulletin Officiel. 2021. Dahir n° 1-21-59 du 3 hija 1442 (14 July 2021) portant promulgation de la loi n° 13-21 relative aux usages licites du cannabis. https://adala.justice.gov.ma/reference/adala_v2/fr/b84cde79-9764-4660-b18d-22ac945751d1.pdf. (accessed 06/11/2023).

CABI. Commonwealth Agricultural Bureau International. Cannabis sativa (hemp). [Online]. https://www.cabi.org/isc/datasheet/14497#toDistributionMaps. Accessed 07/03/2022

Chaachouay N, Azeroual A, Bencharki B, Douira, A, Zidane L. 2022. *Cannabis sativa* L.: A Review on Traditional Uses, Botany, Phytochemistry, and Pharmacological Aspects. Traditional and Integrative Medicine 8: 97-116. https://doi.org/10.18502/tim.v8i1.12407.

Clarke, Robert Connell, and Mark David Merlin. 2013. *Cannabis: evolution and ethnobotany*. Berkeley: University of California Press.

Collins, John. 2020. "A Brief History of Cannabis and the Drug Conventions." *AJIL Unbound* 114: 279-84. https://doi.org/10.1017/aju.2020.55.

Crocq, Marc-Antoine. 2020. "History of Cannabis and the Endocannabinoid System." *Dialogues in Clinical Neuroscience* 22 (3): 223-28. https://doi.org/10.31887/DCNS.2020.22.3/mcrocq.

Dalli M, Azizi SE, Azghar A, Saddari A, Benaissa E, Lahlou YB, Elouennass M, Maleb A. 2023. *Cannabis Sativa* L.: A Comprehensive Review on Legislation, Decriminalization, Phytochemistry, Antimicrobial Activity, and Safety. Journal of Food and Drug Analysis 31 (3): 408-35. https://doi.org/10.38212/2224-6614.3471.

Dr. G.O. Ayenigbara, Dr. G.O. Ayenigbara. 2012. "Medical Utility of Cannabis Sativa." *IOSR Journal of Pharmacy (IOSRPHR)* 2 (3): 460-63. https://doi.org/10.9790/3013-0230460463.

El Khomsi M, Dandani Y, Chaachouay N, Hmouni D. 2022. Ethnobotanical study of plants used for medicinal, cosmetic, and food purposes in the region of Moulay Yacoub, Northeast of Morocco. Journal of Pharmacy & Pharmacognosy Research 10 (1): 13-29. https://doi.org/10.56499/jppres21.1084_10.1.13.

El-Assri EM, El Barnossi A, Chebaibi M, Hmamou A, El Asmi H, Bouia A, Eloutassi N. 2021. Ethnobotanical survey of medicinal plants in Taounate, Pre-Rif of Morocco. Ethnobotany Research and Applications 22. https://doi.org/10.32859/era.22.36.1-23.

El-Hilaly J, Hmammouchi M, Lyoussi B. 2003. Ethnobotanical Studies and Economic Evaluation of Medicinal Plants in Taounate Province (Northern Morocco). Journal of Ethnopharmacology 86 (2-3): 149-58. https://doi.org/10.1016/S0378-8741(03)00012-6.

ElSohly M-A., Radwan M-M, Gul W, Chandra S, Galal A. Phytochemistry of Cannabis Sativa L. Phytocannabinoids. 2017;103:-136. https://doi.org/10.1007/978-3-319-45541-9_1

Farag S, Kayser O. 2017. The Cannabis Plant: Botanical Aspects. In Handbook of Cannabis and Related Pathologies: Biology, Pharmacology, Diagnosis, and Treatment 3-12. https://doi.org/10.1016/B978-0-12-800756-3.00001-6.

Farinon B, Molinari R, Costantini L, Merendino N. The Seed of Industrial Hemp (Cannabis Sativa L.): Nutritional Quality and Potential Functionality for Human Health and Nutrition. Nutrients. 2020;12 (7): 1935. https://doi.org/10.3390/nu12071935.

Friedman J, Yaniv Z, Dafni A, Palewitch D. 1986. A Preliminary Classification of the Healing Potential of Medicinal Plants, Based on a Rational Analysis of an Ethnopharmacological Field Survey among Bedouins in the Negev Desert, Israel. Journal of Ethnopharmacology 16 (2-3): 275-87. https://doi.org/10.1016/0378-8741(86)90094-2.

Gigopulu O, Geskovski N, Stefkov G, Makreski P. 2022. THCA→ THC conversion and its importance for the stability and quality of the herbal material. Macedonian pharmaceutical bulletin, 68 (Suppl 2) 111 - 112. DOI: 10.33320/maced.pharm.bull.2022.68.04.049

Haddou, S, Mounime K, Loukili E, Ou-Yahia D, Hbika A, Yahyaoui Idrissi M, Legssyer A, Lgaz H, Asehraou A, Touzani R, Hammouti B, Chahine A. 2023. Investigating the Biological Activities of Moroccan *Cannabis Sativa* L Seed Extracts: Antimicrobial, Anti-Inflammatory, and Antioxidant Effects with Molecular Docking Analysis. Moroccan

Journal of Chemistry Vol 11: 1116-1136. https://doi.org/10.48317/IMIST.PRSM/MORJCHEM-V11I04.42100.

Hamarneh, Sami. 1972. "PHARMACY IN MEDIEVAL ISLAM AND THE HISTORY OF DRUG ADDICTION". Medical History 16 (3): 226-37. https://doi.org/10.1017/S0025727300017725.

Hartsel, Joshua A., Joshua Eades, Brian Hickory, and Alexandros Makriyannis. 2016. "Cannabis Sativa and Hemp." In *Nutraceuticals*, 735-54. Elsevier. https://doi.org/10.1016/B978-0-12-802147-7.00053-X.

Haut-Commissariat au Plan (HCP). 2023. Les migrants ruraux au Maroc : une confirmation de la féminisation de l'exode et une sélection des jeunes les plus entreprenants. https://www.hcp.ma/file/236167/. (Accessed 22/01/2024)

Hourfane S, Mechqoq H, Bekkali AY, Rocha JM, El Aouad N. 2023. A Comprehensive Review on *Cannabis Sativa* Ethnobotany, Phytochemistry, Molecular Docking and Biological Activities. Plants 12 (6): 1245. https://doi.org/10.3390/plants12061245.

Ibijbijen J, Daoudi A, Bammou M, Zarkani S, Slimani I, Nassiri L. 2016. Ethnobotanical study of medicinal flora in the rural commune of Aguelmouss province of Khénifra (Morocco). Phytotherapie 14: 220-28. https://doi.org/10.1007/s10298-015-0953-z.

Jan HA, Mir TA, Bussmann RW, Jan M, Hanif U, Wali S. 2023. Cross-Cultural Ethnomedicinal Study of the Wild Species of the Genus Berberis Used by the Ethnic Communities Living along Both Sides of the Indo-Pak Border in Kashmir. Ethnobotany Research and Applications 26: 1-14.

Judžentienė A, Garjonytė R, Būdienė J. 2023. Phytochemical Composition and Antioxidant Activity of Various Extracts of Fibre Hemp (*Cannabis Sativa* L.) Cultivated in Lithuania. Molecules 28 (13): 4928. https://doi.org/10.3390/molecules28134928.

Kachmar MR, Naceiri Mrabti H, Bellahmar M, Ouahbi A, Haloui Z, El Badaoui K, Bouyahya A, Chakir S. 2021. Traditional Knowledge of Medicinal Plants Used in the Northeastern Part of Morocco. Evidence-Based Complementary and Alternative Medicine. https://doi.org/10.1155/2021/6002949.

Kaushal, S. (2012). Ethnobotanical study and socioeconomic survey of Cannabis sativa. *Indian Journal of Plant Sciences*, *1*(1), 114-119.

Kim J, Choi P, Park YT, Kim T, Ham J, Kim JC. 2023. The Cannabinoids, CBDA and THCA, Rescue Memory Deficits and Reduce Amyloid-Beta and Tau Pathology in an Alzheimer's Disease-like Mouse Model. International Journal of Molecular Sciences 24 (7): 6827. https://doi.org/10.3390/ijms24076827.

Lapierre, Éliana, Adrian S. Monthony, and Davoud Torkamaneh. 2023. "Genomics-Based Taxonomy to Clarify Cannabis Classification." *Genome* 66 (8): 202-11. https://doi.org/10.1139/gen-2023-0005.

Li, Hui-Lin. 1974. "The Origin and Use of Cannabis in Eastern Asia Linguistic-Cultural Implications." *Economic Botany* 28 (3): 293-301. https://doi.org/10.1007/BF02861426

Liu, Fei-Hu, H.-R Hu, G.-H Du, Gaoju Deng, and Y. Yang. 2017. "Ethnobotanical research on origin, cultivation, distribution and utilization of hemp (Cannabis sativa L.) in China" 16 (April): 235-42.

Mahdizadeh, Shahla, Maryam Khaleghi Ghadiri, and Ali Gorji. 2015. "Avicenna's Canon of Medicine: a review of analgesics and anti-inflammatory substances". *Avicenna Journal of Phytomedicine* 5 (3): 182-202.

Mbuni YM, Wang S, Mwangi BN, Mbari NJ, Musili PM, Walter NO, Hu G, Zhou Y, Wang Q. 2020. Medicinal Plants and Their Traditional Uses in Local Communities around Cherangani Hills, Western Kenya. Plants 9 (3): 331. https://doi.org/10.3390/plants9030331.

McGregor, Iain S., Elizabeth A. Cairns, Sarah Abelev, Rhys Cohen, Mat Henderson, Daniel Couch, Jonathon C. Arnold, and Natalie Gauld. 2020. "Access to cannabidiol without a prescription: A cross-country comparison and analysis". *International Journal of Drug Policy* 85 (November): 102935. https://doi.org/10.1016/j.drugpo.2020.102935.

McPartland, John M. 2018. "Cannabis Systematics at the Levels of Family, Genus, and Species." *Cannabis and Cannabinoid Research* 3 (1): 203-12. https://doi.org/10.1089/can.2018.0039.

McPartland, John M., William Hegman, and Tengwen Long. 2019. "Cannabis in Asia: Its Center of Origin and Early Cultivation, Based on a Synthesis of Subfossil Pollen and Archaeobotanical Studies." *Vegetation History and Archaeobotany* 28 (6): 691-702. https://doi.org/10.1007/s00334-019-00731-8.

Merlin, Mark. 1972. *Man and Marijuana: Some Aspects of their Ancient Relationship.*

Merzouki A, Molero Mesa J. 1999. La Chanvre (*Cannabis sativa* L.) dans La Pharmacopée Traditionnelle du Rif (Nord du Maroc). Ars Pharmaceutica 40 (4): 233-40.

Mikos, Robert A., and Cindy D. Kam. 2019. "Has the 'M' Word Been Framed? Marijuana, Cannabis, and Public Opinion." Edited by Lucy J. Troup. *PLOS ONE* 14 (10): e0224289. https://doi.org/10.1371/journal.pone.0224289.

Mikuriya, T H. 1969. "Marijuana in medicine: past, present and future." *California Medicine* 110 (1): 34-40.

Mohammadi T, Moazzeni H, Pirani A, Vaezi J, Motahhari K, Joharchi MR, Bussmann RW. 2023. Ethnobotany of Plants Used by Indigenous Communities in Birjand, a Dry Region with Rich Local Traditional Knowledge in Eastern Iran. Ethnobotany Research and Applications 26: 1-40.

Nahas, G. G. 1982. "Hashish in Islam 9th to 18th century." *Bulletin of the New York Academy of Medicine* 58 (9): 814-31.

National Academies of Sciences, Engineering, and Medicine. 2017. The Health Effects of Cannabis and Cannabinoids: The Current State of Evidence and Recommendations for Research. Washington, DC: The National Academies Press. https://doi.org/10.17226/24625.

Nawaz, A.H., M. Hossain, M. Karim, M. Khan, Rownak Jahan, and Mohammed Rahmatullah. 2009. "An ethnobotanical survey of Rajshahi district in Rajshahi division,

Bangladesh". *American-Eurasian Journal of Sustainable Agriculture* 3 (January): 143-50.

Oksanen J, Blanchet FG, Kindt R, Legendre P, Minchin P R, O'Hara R B, Simpson GL, Solymos P, Stevenes MHH, Wagner H. 2015. Vegan: community ecology package. R package version 2.5. 6. 1-29.

Osterberger, Elisabeth, Ulrike Lohwasser, Dijana Jovanovic, Joana Ruzicka, and Johannes Novak. 2022. "The Origin of the Genus Cannabis". *Genetic Resources and Crop Evolution* 69 (4): 1439-49. https://doi.org/10.1007/s10722-021-01309-y.

Pagano C, Navarra G, Coppola L, Avilia G, Bifulco M, Laezza C. 2022. Cannabinoids: Therapeutic Use in Clinical Practice. International Journal of Molecular Sciences 23 (6): 3344. https://doi.org/10.3390/ijms23063344.

Pertwee RG. 2006. Cannabinoid pharmacology: the first 66 years. British Journal of Pharmacology 147 (Suppl 1): S163-71. https://doi.org/10.1038/sj.bjp.0706406.

Pisanti, S., and M. Bifulco. 2019. "Medical *Cannabis*: A Plurimillennial History of an Evergreen." *Journal of Cellular Physiology* 234 (6): 8342-51. https://doi.org/10.1002/jcp.27725.

Quarshie, E., & Alagidede, I. P. (2020). Cannabis legalization in Ghana: implications for value-addition in medical and industrial research and applications. *Journal of Indigenous and Shamanic Studies*, *1*(1), 1-14.

R.N, Mary Lynn Mathre. 1997. *Cannabis in Medical Practice: A Legal, Historical and Pharmacological Overview of the Therapeutic Use of Marijuana*. McFarland.

Ren, Guangpeng, Xu Zhang, Ying Li, Kate Ridout, Martha L. Serrano-Serrano, Yongzhi Yang, Ai Liu, et al. 2021. "Large-Scale Whole-Genome Resequencing Unravels the Domestication History of *Cannabis Sativa*". *Science Advances* 7 (29): eabg2286. https://doi.org/10.1126/sciadv.abg2286.

Ross IA. 2005. *Cannabis Sativa*. In Medicinal Plants of the World, Volume 3: Chemical Constituents, Traditional and Modern Medicinal Uses, 29-116. Totowa, NJ: Humana Press. https://doi.org/10.1007/978-1-59259-887-8_2.

Rubin, Vera. 2011. *Cannabis and Culture*. Walter de Gruyter.

Rull, Valentí. 2022. "Origin, early expansion, domestication and anthropogenic diffusion of *Cannabis*, with emphasis on Europe and the Iberian Peninsula". *Perspectives in Plant Ecology, Evolution and Systematics* 55 (June): 125670. https://doi.org/10.1016/j.ppees.2022.125670.

Russo, Ethan B. 2007. "History of Cannabis and Its Preparations in Saga, Science, and Sobriquet". *Chemistry & Biodiversity* 4 (8): 1614-48. https://doi.org/10.1002/cbdv.200790144.

Salmerón-Manzano E, Garrido-Cardenas JA, Manzano-Agugliaro F. 2020. Worldwide Research Trends on Medicinal Plants. International Journal of Environmental Research and Public Health 17 (10): 3376. https://doi.org/10.3390/ijerph17103376.

Schultes, Richard Evans. 1969. "Hallucinogens of Plant Origin". *Science* 163 (3864): 245-54.

Sekkat ZL, Hassikou R, Souad S. 2023. Ethnobotanical study on the use of medicinal plants among diabetic patients in the Rabat-Salé-Kénitra region, Morocco. Ethnobotany Research and Applications, 26, 1-44. - http://dx.doi.org/10.32859/era.71.1-44

Sharma, G. K. 1977. "Cannabis Folklore in the Himalayas". *Botanical Museum Leaflets, Harvard University* 25 (7): 203-15.

Skalli S, Hassikou R, Arahou M. 2019. An ethnobotanical survey of medicinal plants used for diabetes treatment in Rabat, Morocco. Heliyon 5 (3): e01421. https://doi.org/10.1016/j.heliyon.2019.e01421.

Stambouli H, El Bouri A, Bellimam M-A, Bouayoun T, El Karni N. Cultivation of Cannabis Sativa L. in Northern Morocco". Bulletin on Narcotics. 2005;57 (-12): 79118-.

Svrakic, Dragan M., Patrick J. Lustman, Ashok Mallya, Taylor Andrea Lynn, Rhonda Finney, and Neda M. Svrakic. 2012. "Legalization, Decriminalization & Medicinal Use of Cannabis: A Scientific and Public Health Perspective." *Missouri Medicine* 109 (2): 90-98.

Tardío J, Pardo de Santayana M. 2008. Cultural Importance Indices: A Comparative Analysis Based on the Useful Wild Plants of Southern Cantabria (Northern Spain). Economic Botany 62: 24-39. https://doi.org/10.1007/s12231-007-9004-5.

The Angiosperm Phylogeny Group. 2016. "An Update of the Angiosperm Phylogeny Group Classification for the Orders and Families of Flowering Plants: APG IV". *Botanical Journal of the Linnean Society* 181 (1): 1-20. https://doi.org/10.1111/boj.12385.

Thomas B-F, ElSohly M-A. The Analytical Chemistry of Cannabis. 2016. Elsevier. https://doi.org/10.1016/C2014-0-03861-0.

Touw, M. (1981). The religious and medicinal uses of Cannabis in China, India and Tibet. *Journal of psychoactive drugs*, *13*(1), 23-34.

United Nations Office on Drugs and Crime (UNDOC). 2004. Morocco cannabis survey 2004. https://www.unodc.org/pdf/research/Morocco_survey_2004.pdf. (Accessed 17/09/2023).

Vergara, Daniela, Cellene Feathers, Ezra L. Huscher, Ben Holmes, Jacob A. Haas, and Nolan C. Kane. Haas, and Nolan C. Kane. 2021. "Widely Assumed Phenotypic Associations in Cannabis Sativa Lack a Shared Genetic Basis". *PeerJ* 9: e10672. https://doi.org/10.7717/peerj.10672.

Warf, Barney. 2014. "High Points: An Historical Geography of Cannabis." *Geographical Review* 104 (4): 414-38. https://doi.org/10.1111/j.1931-0846.2014.12038.x.

Watts G. Cannabis Confusions. BMJ. 2006;332 (7534): 17576-. https://doi.org/10.1136/bmj.332.7534.175.

Xu Z, Deng M. Cannabaceae. Identification and Control of Common Weeds. 2017;Volume 2, -11924. https://doi.org/10.1007/978-94-024-1157-7_20

Yadav VK, Srivastava A. Microscopical and Chemical Study of Cannabis sativa. Journal of Forensic Research. 2014;05 (01). https://doi.org/10.4172/2157-7145.1000210.

Zhang, Qingying, Xuan Chen, Hongyan Guo, Luisa M. Trindade, Elma M. J. Salentijn, Rong Guo, Mengbi Guo, Yanping Xu, and Ming Yang. 2018. "Latitudinal Adaptation and Genetic Insights Into the Origins of Cannabis sativa L." *Frontiers in Plant Science* 9. https://www.frontiersin.org/journals/plant-science/articles/10.3389/fpls.2018.01876.

More Books!

yes

I want morebooks!

Buy your books fast and straightforward online - at one of world's fastest growing online book stores! Environmentally sound due to Print-on-Demand technologies.

Buy your books online at
www.morebooks.shop

Kaufen Sie Ihre Bücher schnell und unkompliziert online – auf einer der am schnellsten wachsenden Buchhandelsplattformen weltweit! Dank Print-On-Demand umwelt- und ressourcenschonend produzi ert.

Bücher schneller online kaufen
www.morebooks.shop

info@omniscriptum.com
www.omniscriptum.com

OMNIScriptum

MIX
Papier aus verantwortungsvollen Quellen
Paper from responsible sources
FSC® C105338

Printed by Books on Demand GmbH, Norderstedt / Germany